Educating Noncommissioned Officers

A Chronology of Educational Programs for the American Noncommissioned Officer

Daniel K. Elder

All Rights Reserved

Third Edition

May 2020

Title: **Educating Noncommissioned Officers:** A Chronology of Educational Programs for the American Noncommissioned Officer
Author: Daniel K. Elder

NCO Historical Society
4201 W Stan Schlueter Lp, Ste. G
Killeen, TX 76549
www.ncohistory.com

I have tried to credit authors, photographers and owners of prints or photographs where possible and make no claims or credit. We apologize for those instances where it was not.

Cover Photo credit: Edward J. Stuczko, printed with permission from Ed Stuczko.

Contact the NCO Historical Society for changes or updates at info@ncohistory.com.

All photographs are property of the author, unless otherwise noted.

ISBN-13: 978-0-9963181-2-9

Published in the United States of America

Table of Contents

CITATIONS AND CREDIT:
PHOTOS AND IMAGES

Cover photo Courtesy Edward J. Stuczko

Page 6, public domain: US Army

Page 7, public domain: John Lossing Benson

Page 10, public domain: Library of Congress

Page 12, public domain: US Army, Center of Military History

Page 16, public domain: National Archives

Page 17, public domain: Author's collection

Page 19, public domain: US Army, Fort Huachuca Museum

Page 21, public domain: US Army, Infantry Journal

Page 23, public domain: Tom Mendel

Page 26, public domain: National Archives

Page 27, public domain: US Army, Constabulary Headquarters

Page 28, public domain: US Army, Constabulary Headquarters

Page 30, public domain: US Army, Constabulary Headquarters

Page 32, public domain: US Army

Page 33, public domain: US Army

Page 37, Courtesy Edward J. Stuczko

Page 39, public domain: US Army. Author's collection

Page 41, public domain: US Army

Page 43, public domain: US Army, Seventh Army NCO Academy

Page 43, public domain: US Army, Seventh Army NCO Academy

Page 44, Author's collection

Page 46, public domain: US Army

Page 125, public domain: US Army.
Page 127, public domain: US Army.
Page 131, public domain: US Army.

To the best of my ability I have verified that all photographs and images or other visual information that are credited as coming from the Library of Congress, the National Archives, and the Department of Defense and US Army are copyright-free and published by the U.S. Government, or are available from a U.S. government source. Department of Defense (DoD) photographs and imagery, unless otherwise noted, are in the public domain.

The appearance of U.S. Department of Defense visual information does not imply or constitute DoD endorsement.

Introduction

It is too much practice to commit the charge of the elementary drills to non-commissioned officers, by which great many evils are produced… the chance of finding non-commissioned officers who can clearly comprehend and explain the principles of good discipline is not one in twenty.
-William Duane, *Regulations for the Discipline of the Infantry*, 1814

Throughout American history, training noncommissioned officers of the United States Army had been accomplished using on-the-job training (OJT) in the unit, and many believed that is where it should stay. Training of enlisted leaders was conducted by officers in the regiment and was the commanding officer's responsibility. It was accepted that unit training was the best means of developing noncommissioned officers and potential noncommissioned officers.

It was not until the post-World War II era that NCO training was conducted outside the unit at specially designed schools and academies. The first Sergeant Major of the Army, William

1

O. Wooldridge, noted that in those days a soldier had to provide for his own education and training. "I went to night school. There were no requirements to attend school if you did not want to. Now you must get training, or you don't get promoted."

Today, the Army recognizes that NCO leader development is a combination of experiences in operational units, periods of institutional training at specialized NCO schools, and through self-development. Modern noncommissioned officers are held in high esteem in and out of military service, recent policy describes NCOs as "accomplished military professionals who are the Army's preeminent body of leadership."

The NCO corps was not always considered as a professional organization and unlike the officer corps, noncoms did not have a formal system of professional development. Until the post-Vietnam era never had there been a prescribed career pattern or explicit career guidance for NCOs, particularly a system of education and training. Up until the end of the draft in 1972, the Army had benefited from a limitless source of manpower through selective service. The draft brought an abundance of educated men to be trained who filled the ranks and many only served one term of enlistment. After the draft ended the need to develop a career management program became evident and educating NCOs became key to building a professional corps of noncommissioned officers.

At the time of the original writings in 1999 there had been little detailing of the history of noncommissioned officers. In retrospect, I suggest the first significant period of study was after the decree by Secretary of the Army John O. Marsh that 1989 would be known as the first "Year of the NCO." That proclamation energized researchers, authors, and the Army in serious and separate study of the enlisted careerists known as noncommissioned officers. That year the Army's Center of Military History released *"The Story of the Noncommissioned Officer,"* a collection of historical events relating to the history of the "noncom". Until then, in this authors view, the study of the American NCO had been given little consideration.

In an April 1978 response back to retired General Bruce C. Clarke, then Chief of Staff for the Army Training and Doctrine Command Maj. Gen. Robert C. Hixon replied *"Thank you for sending us the letter from Command Sergeant Major Kroger expressing difficulty in uncovering information on the history of the NCO, in general, and on NCO Academies in particular. As a matter of fact, Colonel Karl Morton, first commandant of the Sergeants Major Academy at Fort Bliss [Texas], experienced similar problems with the history of the NCO."* It is the authors intent with this works, and the subsequent updates, to add to that body of knowledge on NCO academy history.

In most early writings the NCO was regarded with the enlisted man and as such received little interest. This updated work attempts to further document one portion of the history leading to the development of career enlisted leaders and their professional development through education; because of the profession of arms education is intertwined with training. I hope to bring together many known sources, to expand on the minimal writing with rediscovered information, and I attempt to correct some long-held misbeliefs in the creation and development of the modern NCO academy.

Updated for the May 2020 edition.
Cmd. Sgt Major Daniel K. Elder, U.S. Army, Retired
csm@danelder.com

Beyond School of the Soldier

They should teach the soldiers of their squads how to dress with a soldier-like air, how to clean their arms, accoutrements, etc. and how to mount and dismount their firelocks.

-Friedrich Wilhelm von Steuben,
"Blue Book" *Instructions for the Sergeant and Corporal*, 1779

One of the earliest mentions of educating noncommissioned officers outside the unit was during the early days of the Continental Army. When Prussian officer Friedrich von Steuben arrived at Valley Forge, he

Baron Friedrich W. von Steuben.
Photo Credit *US Army*

recognized many problems, particularly in discipline, supply, and training. As he set out to restore discipline, Steuben developed tactics using a simple form of manual of arms. In

March 1778 General George Washington had
created the Commander-in-Chief's Guard for
his protection, 100 men of the unit were to
serve as participants in Steuben's experiment,
the development of a new Continental Army.

Illustration of the banner of the Commander-in-Chief's Guard, 1851.
Photo Credit *John Lossing Benson*

Steuben began by drilling one squad, then
allowed sub inspectors to drill other squads
under his supervision. Officers distanced

themselves from the soldiers in the British tradition, but Steuben encouraged them to use sergeants to pass instructions to the drilling troops. Washington was so impressed with the results that he directed that all drilling stops under the current system and that Steuben's simple methods be used.

Though later the officers applied Steuben's techniques to train the soldiers of their troops and regiments, this is probably one of the earliest examples of a specially designed "school" to train both noncommissioned officers and officers outside the unit.

Washington's Army was demobilized four days after the war's formal end, except for 600 men to guard the supplies of the Army. Many Americans felt at the time that a large standing Army in peacetime could infringe on the liberties of the nation. In response, Congress called for states to maintain militias.

By May 1792 the basic militia law was passed which called for the enrollment of "every able-bodied white male citizen between 18 and 45" and the organization of the militia into divisions, brigades, regiments, battalions, and companies by the individual states, each militiamen providing his own arms, munitions, and other accouterments. Those companies, predecessor to the National Guard, met regularly and were trained by elected officers.

The Army had suffered without skilled technicians since the Revolution and many, including General's Washington and Henry Knox, had recommended the development of a military school. In 1802 personnel from the newly created Corps of Engineers were assigned to West Point to serve as the staff for a U.S. Military Academy to teach military science to select officers, and formal military training was introduced to the Army.

Secretary of War John C. Calhoun proposed the first specialist school in 1824, in that a "school of practice" be established, from which the Artillery School at Fortress Monroe was developed. Unlike modern schools which taught individuals, this school taught entire

Illustration of Fortress Monroe, 1861.
Photo Credit *Library of Congress*

units, including enlisted men. It was closed 11 years later in 1835 when the students were sent to Florida in response to the Seminole threat and reopened in 1858. By the mid 1870's the school was training noncommissioned officers in the history of the United States, geography, reading, writing, and mathematics.

In 1868, the Signal Corps established a Signal School which was opened at Fort Greble, DC and, in 1869, moved to Fort Whipple, VA (later Fort Myer). By 1871 the school at Whipple had a primary duty to train observer-sergeants and assistant observers in their duties. Candidates were selected from the enlisted men of the signal detachment and they were taught technical subjects on military signaling, telegraphy, and meteorology and pursued a regular course of study. Leadership was typically not part of the curriculum.

Signal School, circa 1920.
Photo Credit *US Army*

The candidates were promoted to observer-sergeant after completing their studies, six-months of practical application and after appearing before an examination board of officers. Similar to the Artillery and Signal Schools, other technical schools proliferated.

Other technical schools included the School of Application for Infantry and Cavalry at Fort Leavenworth, the Engineer School of

Application at Fort Totten, New York and the Army Medical School were also established in the mid to late 1800's. Most of the classes were conducted indoors, except for the less technical courses, such as infantry and cavalry.

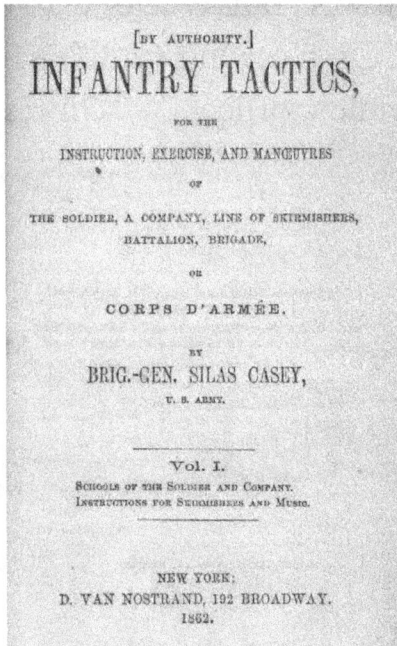

[BY AUTHORITY.]

INFANTRY TACTICS,

FOR THE

INSTRUCTION, EXERCISE, AND MANŒUVRES

OF

THE SOLDIER, A COMPANY, LINE OF SKIRMISHERS, BATTALION, BRIGADE,

OR

CORPS D'ARMÉE.

BY

BRIG.-GEN. SILAS CASEY,
U. S. ARMY.

Vol. I.
SCHOOLS OF THE SOLDIER AND COMPANY.
INSTRUCTIONS FOR SKIRMISHERS AND MUSIC.

NEW YORK:
D. VAN NOSTRAND, 192 BROADWAY.
1862.

Casey's Infantry Tactics manual.

Around the time of the Civil War Maj. Gen. Silas Casey called for NCO education in his book on tactics, insisting that NCOs be formally trained to give commands on the battlefield. But he had to overcome the opposition of company grade officers. They argued that company commanders knew their men's capabilities and limitations best and were in a better position to provide them on-the-job training.

A minority of officers doubted that OJT could meet the needs for the combat arms and wanted more post schools. But World War I would begin with NCOs receiving traditional unit instruction, while officers' schools multiplied.

Noncommissioned Officer Schools

The noncommissioned officer are men who in civil life would be skilled workmen, foremen, chief clerks and subordinate officials. If the Army can not offer them inducements that civil institutions are glad to offer, it cannot hope to secure or retain them.

-War Department Annual Report (Vol. I),
Washington, D.C., 1907, p. 83

It would take the casualties of war to answer the need for noncommissioned officer leadership schools. By the early summer of 1918, the United States had been at war in Europe for over a year. Training camps and recruit depots were established to develop replacements for the force overseas, American NCOs were viewed only as half-trained and unsophisticated by allied noncoms. Detachments of US troops were shipped off to Europe as soon as they could be inducted, clothed, equipped and minimally trained.

There was no reserve of soldiers and sergeants for which the Army could draw from and there was no method to rapidly develop replacement noncommissioned officers for Gen. John J.

Pershing's American Expeditionary Force. Because of the shortage of NCOs, the Army Staff recalled 648 retired men to serve as recruit trainers.

Soldiers marching at Camp Dix, NJ, 1918.
Photo Credit *National Archives*

These older noncoms at the training centers were needed to train replacements, so noncommissioned officers were selected and designated from within the ranks. Initially there were no noncommissioned officer training schools as a result unless company officers could find the time for additional instruction

these new NCOs only received basic
instructions for an infantry soldier.

AT THE SCHOOL FOR NON-COMMISSIONED OFFICERS AT CAMP MEADE, MARYLAND, WHERE THEY ARE LEARNING TO FIRE FROM THE TRENCHES UNDER THE SUPERVISION OF BRITISH OFFICER-INSTRUCTORS. (© International Film Service.)

Training at a Camp Meade, MD WWI-era NCO School.
Author's collection

The results were evident in unacceptably high
casualty lists of all ranks. Pershing called out
that "more stress be laid upon the
responsibility in the training of sergeants. They
will be imbibed with the habit of command and
will be given schooling and prestige to enable

them to replace officers once casualties." The Secretary of War directed that "their [noncoms'] duties and responsibilities will be thoroughly represented to them, by means of school courses and official [interaction] with their immediate commanding officer."

The War Department responded by issuing a directive that required out of each detachment of replacements that a "sufficient number of men be selected, segregated, and especially trained as noncommissioned officers." This would result in the development of eleven hundred corporals and sergeants every thirty-days of training. Though not institutionalized, a similar method of training replacement NCOs would be adopted for a different war in a different time. For WWI, it served as the "next best" means to secure large numbers of noncommissioned officers for the war effort.

Once instituted, raw, untrained men were handpicked from among their fellow soldiers and put through a coordinated and standardized program of instruction. This was conducted and supervised by specially selected officers trained for this purpose. Though these graduates were only slightly more competent than their peers who had only completed basic military training, they were an improvement over the alternative.

25th Infantry Regiment bayonet practice, 1928.
Photo Credit *US Army*

The results did exceed expectations and the centralized training program spread to other replacement camps. In 1919 the 25th Infantry Regiment established such a school and they recognized a higher standing of efficiency among their noncommissioned officers. Although a call for continuation of this noncommissioned officer training program through a regimental noncommissioned officer school went out, it would not be until later years that NCO leadership schools would return.

Pershing's expedient would not survive demobilization. But the selection and training of noncommissioned officers would continue as a subject of much debate.

Post-World War II and Occupation Duty

No two people will agree on how to train NCO's. The book, however, says that training is a command responsibility, and this doctrine must be observed.

Capt. Mark M. Boatner,
School for Noncoms, Infantry Journal, 1947

Prior to World War II some regiments and divisions had established NCO schools, but as the war progressed, combat attrition rapidly thinned the ranks of these trained noncoms. Hastily devised training programs for WWII produced NCOs primarily trained to fight rather than lead soldiers in a garrison. New inductees would receive hands-on instruction for their basic combat training, then were sent to their unit for additional training. A man was promoted if he showed potential, with privates becoming corporals, and corporals, sergeants.

These NCOs were not as capable as their pre-war predecessors, and the experience level of the noncommissioned officer continued to decrease. By the war's end rapid demobilization and high personnel turbulence conflicted with the Army's role for occupation duty in Europe. Many of the replacements sent overseas after the war had little training or combat experience and the Army was weakened by a shortage of good noncommissioned officers.

Inspection Time, Blue Devil style.
Photo Credit *US Army*

Realizing the need to prepare soldiers for the specialty duties required of the occupation trooper, the 88th Infantry Division established its own training center. The "Blue Devils," on duty in Venezie Giulia, Italy set out to develop a more professional noncommissioned officer, and the division established the Lido Training Center on Lido Island near Venice in November 1945. The training center was not a school in the traditional sense, but a model battalion not unlike the Fortress Monroe Artillery School in which the noncoms lived "by the book" for six weeks.

The program was built around teaching discipline and enforcement of standards, and also taught leadership, guard orders, customs and courtesies, and other typical duties expected of NCOs.

Trieste United States Troops (TRUST) Insignia.
Photo Credit *Tom Mendel*

Inspections, physical fitness training and close order drill were conducted daily, and instantaneous obedience to orders was expected. Most of the instructors at the center were corporals and sergeants instead of officers. Peers and instructors evaluated each of the students as they performed one of the 130 different jobs at the school.

The graduating students showed enthusiasm for the program and the chance to learn by actually *doing*, and commanders were pleased with the results. By mid-1947 the Center had trained almost 4,000 students, but the 88th would return to the United States later in that same year and discontinue its school.

The Constabulary School

Training difficulties arose (at the Constabulary School) because of the shortage of instructors and the lack of appropriate texts to issue to the students.

-Capt. Dee W. Pettigrew, Historian, U.S. Constabulary School, July 1946

To replace the inactivating divisions on occupation duty the United States European Command organized the United States Constabulary. Heavily armed, lightly armored, and highly mobile, the Constabulary were enforcers of law, support to authorities and would serve as a covering force in the event of renewed hostilities. In January 1946, the Third U.S. Army Commander, Lt. Gen. Lucian Truscott, gave the task of organizing this force to Maj. Gen. Ernest Harmon. Harmon was given until July to have this force readied to carry out its assigned tasks and would be headquartered in Bamberg. Early in the planning stages the need for a Constabulary school became evident. The Constabulary trooper needed to not only know the customary duties of a soldier, but police methods, how to make arrests, and how to deal with the local population.

Students undergoing an inspection by Constabulary officers.
Photo Credit *National Archives*

The return of units, divisions, and skilled combat veterans to the United States had plagued the theater with an abundance of minimally trained and unhappy soldiers. The majority of military personnel in Europe were re-enlistees or freshly inducted troops, with some lacking even the most basic training.

The 1st and 4th Armored Divisions were selected as the nucleus to form the Constabulary and Harmon set out to instill a Constabulary spirit that would reflect the pride and importance of their duties. Harmon directed that a school be established, and Col. Harold G. Holt was selected as the first Commandant. A group of training cadre instructors was assembled in Bad Tölz, and Harmon outlined the mission of the school, the subjects to be taught, and the

The Constabulary School in Sonthofen.
Photo Credit *US Army*

standards that would be met.

Maj. Gen Ernest N. Harmon.
Photo Credit *US Army*

In February, the former *Adolf Hitler Schule*, located at Sonthofen, was selected as the site for the Constabulary school. The 2d Cavalry Squadron began preparation for the school's early operation and was replaced in February by the 465th Anti-Aircraft Automatic Weapons Battalion, redesigned the Academic Troop, Constabulary School Squadron. By March 4, 1946, the first class of 129 officers and 403 enlisted men reported to Sonthofen.

Harmon explained the need for training on

graduation day to this first class that:

> *The Constabulary School is more than a place of instruction. It is a cradle, so to speak, in which we hope to establish the character, the espirit de corps, high standards of personnel [sic] conduct, and appearance of the Constabulary. As most subjects taught here are entirely new to the soldier and the normal training of soldiers, it was felt necessary to obtain as quickly as possible the maximum number of graduates to act as instructors to their units and to spread the Constabulary standards.*

The decision was made after the first course was completed to separate the officer and enlisted students to devote specific training hours to each group. By 1947, every month special trains began at the extreme end of the U.S. Zone heading towards Sonthofen and picked up students along the way.

In January 1947 the 7719th Theater School
(Special) was
consolidated at
Sonthofen and
the Constabulary
School began to
lose its identity.
By then the
Constabulary
School offered
five courses, at

Reporting in at the Constabulary NCO Academy
Photo Credit *US Army*

its core were the basic and advance
Constabulary Courses for enlisted men, and
three non-Constabulary programs, including a
Noncommissioned Officer course. Later, the
Sonthofen school offered courses called for by
War Department Circular No. 9.

The theater-wide NCO Academy was designed
to train NCOs and potential NCOs in their

basic duties and was established at the school on June 30, 1947. The course emphasized basic subjects, supply, and administration. The school trained students from around the theater, not only from the Constabulary, but also from the European Command and Trieste, Italy. Besides the NCO basic and enlisted man's courses, the school also taught a Sergeants Major and First Sergeants course. In mid-1948 the school was closed, and the now defunct school became the headquarters for a Field Artillery Group.

Maj. Gen. Isaac D. White coming from the Calvary School at Fort Riley, Kansas assumed command of the US Constabulary in May 1948 and led a reorganization to convert Constabulary brigades to armored cavalry regiments. White had been tasked to transition the Constabulary from a police force to a "multitasked, hard-hitting armored Cavalry fighting force as soon as possible."

Tensions between the occupying nations in divider Germany were heating up and on June 24, 1948 the Soviet Union severed the connections between the non-Soviet sectors and Berlin. General Lucius D. Clay, in charge of the US Occupation Zone and his forces,

Universal Military Training Experimental Unit at Fort Knox, 1948
Photo Credit *US Army*

were cut off. The Berlin Blockade was to be one of the first major international crises of the

Cold War and would not be lifted until May 12, 1949 without a restart of fighting in the region.

Brig. Gen. Bruce C. Clarke.
Photo Credit *US Army*

In early 1949 the Armor School at Fort Knox, KY, had as its Assistant Commandant Brig. Gen. Bruce C. Clarke. It was Clarke, of St. Vith fame in World War II who went about improving training for the armored force. One of the courses established during his tenure was the Armor School's Noncommissioned Officers Course. Initially, this four-month course was considered the most comprehensive instruction ever presented to noncommissioned officers.

The course was taught by the school's academic groups, employing methods of instruction based on lectures, conferences, demonstrations, and practical exercises. In many cases the students were taught in the classroom on a subject, then conducted a practical exercise to actually use the knowledge "hands-on." In this way the lesson was presented, demonstrated, practiced, and critiqued all in the same day.

SFC Philip Wharton and SGT Frank Mangin, graduates of Armor NCO Course No. 1, 1949 noted that,

> *"This new Army of ours is a group of officers and men with a professional interest in their careers; and these men know the part the noncommissioned officers play is extremely important."*

Students typically received instruction on

leadership, tactics, command and staff, automotive principles, personnel management, among other subjects. Besides training in physical fitness, guest speakers were used to teach the students.

But by mid-1949, the Noncommissioned Officers Course was renamed the Tank Commanders Course and reduced to 13 weeks and was only available to NCOs in Grades 2 and 3 (in 1949 the Career Compensation Act reversed the grade structure) with duties as tank commanders. Though the renamed course was similar in nature to the NCO Course, only noncoms in the Armored Cavalry force would attend.

The Second Constabulary NCO Academy

We propose, in carrying out the academy's primary aim of developing you as leaders? to teach you how to teach others? how to reproduce for your men, the subject matter which you are taught here.

-Brig. Gen. Bruce C. Clarke, Commandant
to first graduating class, 1949

Army Ground Forces became the Army Field Forces (AFF) in 1948 and the following year Maj. Gen. White determined that more specialized training was needed for the noncommissioned officers of the Constabulary. The 2nd Cavalry Brigade was re-designated in July 1949 as the 2d Constabulary Brigade and stationed in Munich. White gave the newly assigned commander Brig. Gen. Clarke the mission of organizing a Noncommissioned Officer Academy in unused buildings at Jensen Barracks in Munich, and to also serve as the Academy's Commandant. Clarke was enthusiastic about the project.

White explained to Clarke what he wanted of the curriculum and stated it would be run on a strict military basis. It was to be purely academic classroom instruction, not hands-on training. Clarke set up a six weeks course with White's approval, and in September 1949 the Constabulary Noncommissioned Officer Academy was established. In later years, Clarke, who would go on to earn four stars, would consider the NCO Academy to be one of the most successful activities he had been charged with in his illustrious career.

Sign at the gate of Jensen Barracks, 1950s.
Photo Credit *Edward J. Stuczko*

Clarke created a staff partly from the officers who had worked under him as students and instructors at the Armored School. Fourteen subjects were decided on to form the basic curriculum and included drill and command, military justice, physical fitness, and basic tactics. As in the 88th Division's school, Clarke's academy required rigid discipline. The three major departments, Leadership and Command, Tactics, and Personnel and Administration were charged with the conduct of the training.

The students' day began at 5 a.m. and continued until taps played at 11 p.m. Soldiers in the first three grades who were not previously officers or graduates of similar training were considered for attendance at this NCO Academy. As with its predecessor in Sonthofen, the Munich NCO Academy was originally established for Constabulary troopers, but the graduates' success spilled over to the other units and soon expanded to the Seventh Army and the European and Trieste Commands.

Col. William A Rau presenting award to distinguished graduate of the Seventh Army NCO Academy.
Photo Credit *US Army*

On October 15, 1949 the first class of 150 students reported to the Constabulary Academy. In later classes the Academy reached their full student load of 320 and by 1951 had graduated almost 4,500 students. As part of developing future noncommissioned officer replacements the Academy allowed enlisted soldiers from Grades 4 and 5 (corporal and private first class) to attend, providing they had the appropriate educational background and demonstrated potential to become a noncom.

To serve as an inspiration to all on the campus, the ten buildings on Jensen Barracks were all named after World War II Medal of Honor recipients who gave their lives in the European Theater. One sergeant who graduated from the Constabulary NCO academy as the number-one student, Sergeant Leon L. Van Autreve, would make history in becoming the first non-combat arms Sergeant Major of the Army (SMA).

Soon after reporting to his engineer company in Boeblingen, Germany, the Belgian-borne Van Autreve's first sergeant became upset that few NCOs volunteered to attend the NCO academy. So, Van Autreve

Sgt. Maj. of the Army Leon L. Van Autreve.
Photo Credit *US Army*

volunteered to attend and upon graduation was rewarded with promotion to E-7, the highest enlisted pay grade at that time.

In later years, after the creation of the "supergrades" (pay grades E-8 and E-9) in 1958, Van Autreve's commander at the Continental Army Command (CONARC) Armor Board was General Clarke. Clarke

41

recommended Van Autreve for the Engineer section's only E-8 position because of his successes at the Constabulary NCO Academy.

Not all noncommissioned officers believed in the importance of NCO education. Platoon sergeant William O. Wooldridge, who would later be selected as the first Sgt. Major of the Army asked his first sergeant for permission to attend the Seventh Army NCO Academy.

Wooldridge explained that he intended to stay in the Army, and "wanted to be something more than a rifle platoon sergeant." His first sergeant scolded that "You're a combat veteran. You already know everything." When Wooldridge continued to press the issue, the first sergeant told him, "you're wasting my time," and ordered him "out of my orderly room."

Standardizing NCO Academies

The purpose of Noncommissioned Officer Academies is to broaden the professional knowledge of the noncommissioned officer and instill in him the self-confidence and sense of responsibility required to make him a capable leader of men.

-Army Regulation 350-90, *Noncommissioned Officer Academies*, June 1957

Seventh Army NCO Academy at Bad Tölz, GE.
Photo Credit *US Army*

In 1951 the Seventh Army assumed the Constabulary functions and the Constabulary NCO Academy became the Seventh Army Noncommissioned Officers Academy, and in November 1958, moved to Flint Kaserne in Bad Tölz. Some 45,000 noncoms had graduated from the Munich school by then, with even several students from the newly formed West German *Bundeswehr* attending. Clarke would go on to establish other NCO Academies in Texas, Hawaii, and Korea, and other divisions began to develop their own versions. The general purpose of these academies was "to teach noncommissioned officers to look, act, and think like, and accept the responsibilities of noncommissioned officers."

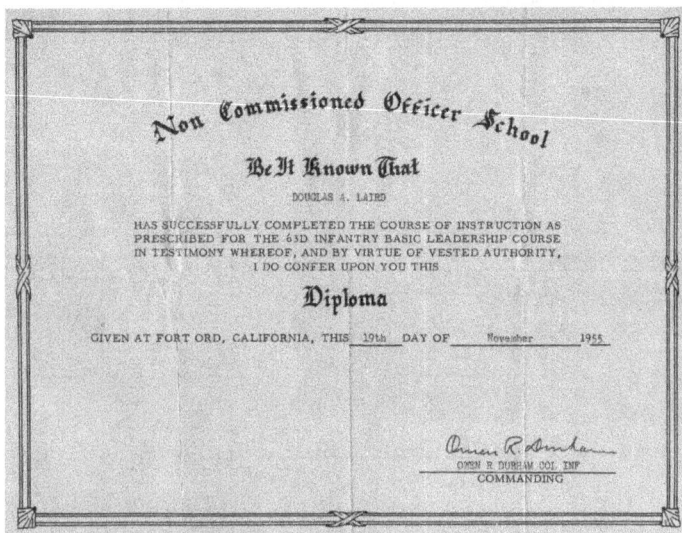

NCO Academy diploma, 1955.
Author's collection

Though they were similar in nature and conduct, there were no established standards of instruction, and graduates of one course could later ultimately be required to attend another. Most noncommissioned officers never attended NCO academies and continued to learn from the old methods of OJT. The Korean War brought an urgent need for better-trained small unit leaders.

At Clarke's urging, the Army's NCO Academy system was developed in 1957 when the Department of the Army published its first regulation to establish standards for NCO Academies.

Instructor at the 2nd Armored Division NCOA.
Photo Credit *U.S. Army*

This June 25th directive stated that the "purpose of Noncommissioned Officer Academies is to broaden the professional knowledge of the noncommissioned officer and instill in him the self-confidence and sense of responsibility required to make him a capable leader of men."

4th Armored Division NCO Academy, Neu Ulm, GE.
Photo Credit *US Army*

The hope was that a better-trained NCO would be needed for the new Pentomic organization of the new cold war era. This regulation authorized, but did not require, division and installation commanders to establish NCO academies. It set forth a standard pattern for training NCOs and fixed the minimum length of a course at four weeks. It did not call for a standardized course of instruction but mandated seven subjects that were required as part of the curriculum and would emphasize the new concepts of atomic warfare.

The regulation required that each command to support its academy from its available resources and did not provide additional funding. For the first time, Army noncommissioned officers had an Army-sponsored program for institutional training. Though the Korean War would derail many programs, noncommissioned officer academies would continue to operate throughout the war.

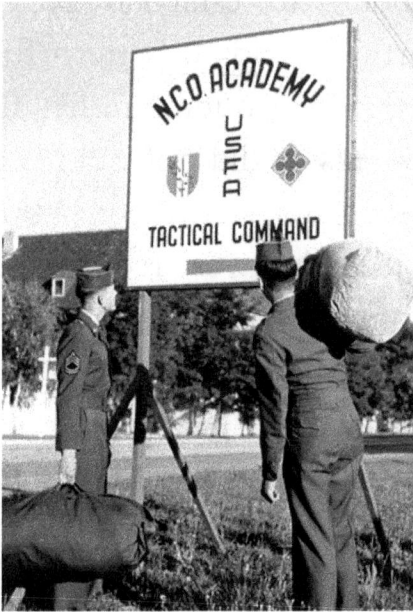

US Forces Austria NCO
Academy sign.
Photo Credit *US Army*

The U.S. occupied post-World War II Austria from 1945 to 1955 and formed the United States Forces in Austria (USFA) with Gen Mark W. Clark as Commanding General. The American occupation forces in Austria operated as a semi-independent command under Headquarters, USFET, until March 1947. After, USFA came under Headquarters, European Command. The USFA Tactical Command established its own NCO Academy in Building 14 on Camp McCauley near Linz, Austria.

With a mission to develop and perfect the "high standards of responsibility, leadership, and instructor ability required in the noncommissioned officers of USFA," its curriculum included subjects that were deemed essential to organizational training and "the science of military tactics and those subjects directly contributing to military proficiency. where it remained until the USFA." With the signing of the Austrian State Treaty in Vienna, the USFA mission ended and Austria was free of occupying troops by October 1955.

Directions to the Seventh Army NCO Academy.
Photo Credit *US Army*

Studying the Effects of NCO Training

The study should answer the questions: What changes occur in a man as a result of Academy attendance? What practices of the Academy seem most important in producing those changes? The above studies would be profitable in considering the kinds of training indicated for NCO Academies in general... .

> -Dr. Francis Palmer, HumRRO, trip report after
> visiting the Seventh Army NCO Academy,
> November 1956

In 1957, CONARC and the U.S. Army Leadership Human Research Unit (under the George Washington University) began to study the feasibility of identifying and training enlisted soldiers in the event of hostilities to perform in leadership roles. Long recognized that the NCO was important to the smooth operation of the Army, there had been relatively little research conducted on improving their training. Task NCO was thus born, with the goal of determining how to identify and train enlisted soldiers as NCOs.

Leadership Human Research Unit
Photo Credit *US Army*

The Human Resources Research Office (HumRRO) of George Washington University developing initial psychological predictors of leadership potential and the evaluation system for use in identifying competent leaders for senior NCO positions.

The U.S. Army Personnel Research Office (USAPRO) in parallel research studies had the mission of developing techniques to identify early in the careers of those enlisted men who might be capable of becoming good

51

noncommissioned officers in the combat arms branches.

The existing Noncommissioned Officer Academy system was selected to serve as the framework to measure leadership performance. Initially, the HumRRO project was to study the effects of academy training on noncommissioned officers job performance and to study the factors that modify effects of academy training. But at the urging of the CONARC Human Research Advisory Committee, HumRRO ultimately settled on a three-phase study. The research was begun under the name of Task NCO, but near the end of the studies was designated Work Unit NCO.

These research projects were divided into 3 major phases, dubbed NCO I, NCO II and NCO III (two phases). Under NCO I the Army's training system for enlisted personnel and its methods for selecting and training NCOs was examined. As a result, USCONARC Pamphlet 350-24, *Guide for the Potential Noncommissioned Officer,* was published. In the midst of these studies and field experiments, the Department of Defense was faced with a possible call for mobilization during the Berlin Crisis in 1961. The pressure was on to develop predictors to identify potential NCOs.

HumRRO suggested that a two-week Leader Preparation Course (LPC) between basic and advanced training be instituted. The goal was to provide support to the training cadre at advanced training sites and centers and provide these trainee-leaders with supervisory and human relations skills. In October, a Leader Preparation Program (LPP) was implemented at Forts Carson, Dix, Gordon, Jackson, Knox, and Ord. In 1963 a one-week Leader

Orientation Course was provided to the Women's Army Corps, to be conducted one week before basic training.

The LPP was based on a four-week LPC and consisted of training programs and observations.

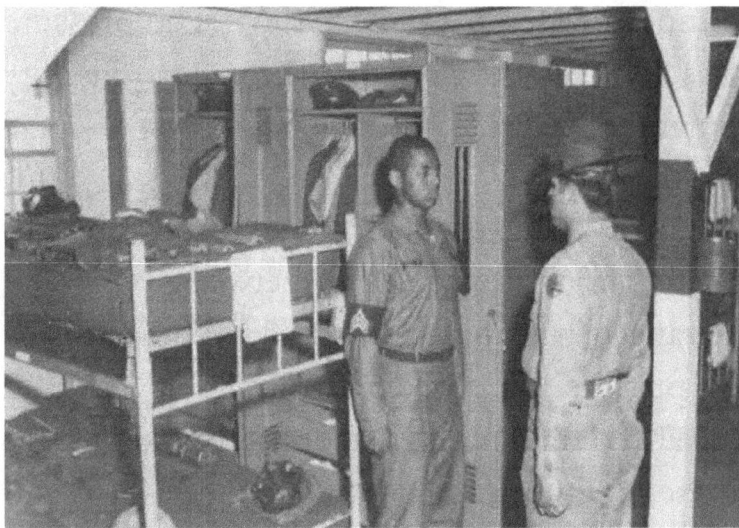

Clothing and Equipment Inspection of Trainee leader at Fort Ord, CA.
Photo Credit *US Army*

At Fort Ord, CA the Leadership Academy

served as the Training Command's Leadership and Professionalism Training Group. It had the Drill Sergeant School and the Sixth Army Noncommissioned Officer School and was responsible for the Leader Preparation Course, an instructor training course, and the basic leader's course prescribed by the Department of the Army. The Ft. Ord Leader Preparation Program identified basic combat trainees who showed high leadership potential and gave them leadership training at the leadership course early in the training cycle. That way they could gain leadership skills earlier and then employ their knowledge in leadership positions during the remainder of basic combat training. Typically, twenty-five men from a basic combat training company were chosen with no loss of morale noted among those not selected for the program.

As to be expected for a program of this type,

there was resistance from the "old soldiers." The researchers noted that at each training center they had to contact and persuade approximately 30 officers and 100 NCOs to adjust their procedures, and to convince them the process to identify potential leaders would work.

Third Army Noncommissioned Officer Academy, Oct. 1964.
Photo Credit *US Army*

It was during NCO II that a series of pilot studies was conducted to examine the problems of junior NCO selection, prediction and evaluation of new recruits. Informal leadership training was conducted using different approaches and techniques, and by the completion of NCO III, three experimental training systems were developed. The conclusions drawn up in 1967 at the close of their 10-year study on how to train NCOs and potential NCOs were:

Leadership Selection. The candidate for leadership training should be above average on BCT (basic training) Peer Ratings and on the appropriate Aptitude Area score. Supervisors' evaluations should be used to eliminate men who are obvious misfits or to recommend men who are outstanding prospects in the opinions of the cadre despite poor aptitude scores or low Peer Ratings.

Leadership Training. The experimental training methods led to better leadership indications on nearly all criteria, with the Leader Preparation

Course system exhibiting greatest effectiveness and feasibility among various experimental and control conditions tested.

Training Method. Relatively little criterion difference was found between results from specific training methods (i.e., functional context versus traditional; high cost versus low cost). However, because the time involved in presentation of each different method varied, definitive comparisons could not be made.

A note of interest about the Unit NCO project was the 1964 changes made to basic training at the direction of Secretary of the Army, Steven Ailes. CONARC developed a new concept to transfer responsibility from training committees to the platoon sergeant. Technical advisory in the development of the "Drill Sergeant" was provided by the Work Unit and the LPP served as the model for the Drill Sergeant Program and in developing the Drill Sergeant Course, first conducted at Fort Jackson, SC.

The Noncommissioned Officer's Candidate Course

A strained voice shatters the stillness: Pass in Review. *And at this moment he knows. This is command reveille.* Right Turn March. *It is to characterize the next 12 frustrating weeks of training.*
-Infantry NCO Candidate Course,
Class 4-69 yearbook

By the early-1960's, the United States Army was again engaged in conflict, now in Vietnam. As the war progressed, the attrition of combat, the 12-month tour limit in Vietnam, separations of senior noncommissioned officers and the 25-month stateside stabilization policy began to take its toll to the point of crisis. Without a call up of the reserve forces, Vietnam became the Regular Army's war, fought by junior leaders. The Army was faced with sending career noncoms back into action sooner or filling the ranks with the most senior PFC or specialist.

Field commanders were challenged with understaffed vacancies at base camps, filling various key leadership positions, and providing for replacements. Older and more experienced NCOs, some World War II veterans, were strained by the physical requirements of the methods of jungle fighting. The Army was quickly running out of noncommissioned officers in the combat specialties.

In order to meet these unprecedented requirements for NCO leaders the Army developed a solution, an NCO and Specialist procurement program called the Skill Development Base (SDB). Modeled after the proven Officer Candidate School (OCS) where an enlisted man could attend basic and advanced training, and if recommended or applied for, filled out an application and attended specialized training to be a platoon commander, the thought was the same could be done for noncoms. If a carefully selected soldier can be given 23-weeks of intensive training that would qualify him to lead a platoon, then others can be trained to lead squads and fire teams using similar methods.

From this seed the Noncommissioned Officers Candidate Course (NCOCC) was born. Potential candidates were selected from groups of initial entry soldiers who had a security clearance of confidential, an infantry score of 100 or over, and demonstrated leadership potential. Based on recommendations, the unit commander would select potential NCOs, but all were not volunteers.

Those selected to attend NCOCC were immediately made corporals and later promoted to sergeant upon graduation from phase one. The select few who graduated with honors would be promoted to staff sergeant. The outstanding graduate of the first class, Staff Sgt. Melvin C. Leverick, recalled "I think that those who graduated [from the NCOCC] were much better prepared for some of the problems that would arise in Vietnam."

The NCO candidate course was designed to maximize the two-year tour of the enlisted draftee. The Army Chief of Staff General Harold K. Johnson approved the concept on June 22, 1967, and on September 5 the first course at Fort Benning, GA began. By combining the amount of time it took to attend basic and advanced training, including leave and travel time, and then add a 12-month tour in Vietnam, the developers settled on a 21-22 week program.

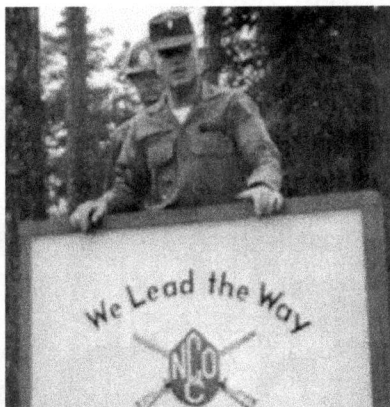

Cadre critiques candidates.
Photo Credit *US Army*

NCOCC was divided into two phases. Phase I was 12 weeks of intensive, hands-on training, broken down into three basic phases. For the Infantry noncom, the course included tasks such as physical training, hand-to-hand combat, weapons, first aid, map reading, communications, and indirect fire. Vietnam veterans or Rangers taught many of the classes, but the cadre of the first course were commissioned officers. The second basic phase focused on instruction of fire team, squad, and platoon tactics.

Though over 300-hours of instruction was given, 80-percent was conducted in the field. The final basic phase was a "dress rehearsal for Vietnam," a full week of patrols, ambush, defensive perimeters, and navigation. Twice daily the Vietnam-schooled Rangers critiqued the candidates and all training was conducted tactically. Throughout the 12-weeks of training, leadership was instilled in all that the students would do. A student chain of command was set up and "Tactical NCOs" or TACs, supervised the daily performance of the candidates.

By the time the students successfully completed Phase I NCOCC, they promoted to sergeant or staff sergeant, and shipped off to conduct a 9-10-week practical

application of their leadership skills by serving as assistant leaders in a training center or unit. This gave the candidate the opportunity to gain more confidence in leading soldiers.

Student Leader NCOCC, Ft Benning.
Photo Credit *US Army*

As with many programs of its time, NCOCC was originally developed to meet the needs of the combat arms. With the success of the course, it was extended to other career fields, and the program became known as the Skill Development Base Program. The Armored School began NCOCC on December 5, 1967. Some schools later offered a correspondence "preparatory course" for those who anticipated attending NCOCC or had not benefited from such formal military schooling.

As with the Leadership Preparation Course tested by HumRRO, the "regular" noncoms and soldiers had much resentment for the NCOCC graduates, as those who took 4-6 years to earn their stripes the hard way, were immediately angered. Old-time sergeants began to use terms like "Shake 'n' Bake," "Instant NCO," or "Whip-n-Chills" to identify this new type on noncom. Many complained by voice or in writing that it took years to build a noncommissioned officer and that the program was wrong.

Many feared it would affect their promotion opportunities, and one senior NCO worried that "nobody had shown them [NCOCC graduates] how to keep floor buffers operational in garrison." William O. Wooldridge, serving as the recently established position of Sergeant Major of the Army had received correspondence and concern that men in Vietnam with the same service, plus combat experience, may not be promoted. He stated in a message to the field that, "promotions given to men who complete the course will not directly affect the promotion possibilities of other deserving soldiers in Vietnam or other parts of the world."

TAC NCO shows Candidates an M60 machine gun components.
Photo Credit *US Army*

In his speech to the first graduating class

Wooldridge said that, "Great things are expected from you. Besides being the first class, you are also the first group who has ever been trained this way. It has been a whole new idea in training." As the Sgt. Maj. of the Army expressed, all were not suspicious of this new way to train NCOs.

Sgt. Maj. Of the Army William O. Wooldridge with Army Ch. of Staff Harold K. Johnson in the Pentagon.
Photo Credit *US Army*

After initial skepticism, one former battalion

commander Col. W. G. Skelton explained, "within a short time they [NCOCC graduates] proved themselves completely and we were crying for more. Because of their training, they repeatedly surpassed the soldier who had risen from the ranks in combat and provided the quality of leadership at the squad and platoon level which is essential in the type of fighting we are doing."

NCOCC graduates recognized the value of their training. Young draftees attending initial training at the time knew they were destined for Vietnam. Many potential candidates were eligible for Officer Candidate School but rejected it because they would incur an additional service obligation and have to spend more time in the Army. They realized that NCOCC was a method by which they could expand on their military training before entering the war.

Some were exposed to the Phase II NCO Candidates serving as TAC NCOs during their initial training and felt they could do the same. Many graduates would later say that the NCO school, taught by Vietnam veterans who experienced the war firsthand, was what kept them and their soldiers alive and its lessons would go on to serve them well later in life. Many were assigned as assistant fire team leaders upon arrival in Vietnam and then rapidly advanced to squad or platoon sergeants.

Most would not see their fellow classmates again, and in many cases were the senior (or only) NCO in the platoon. Some would go on to make a career of the military or later attend OCS, and four were Medal of Honor winners. In the end almost 33,000 soldiers were graduates of one of the NCO Candidate

A graduation of the Ft. Benning NCO Candidate Course.
Photo Credit *US Army*

Courses.

The NCOCC graduate had a specific role in the Army-they were trained to do one thing in one branch in one place in the world, and that was to be a fire team leader in Vietnam. It was recognized that they were not taught how to teach drill and ceremonies, inspect a barracks, or how to conduct police call. Many rated the program by how the graduates performed in garrison, for which they had little skill. But their performance in the rice paddies and jungles as combat leaders was where they took their final tests, of which many receiving the ultimate failing grade. But educating NCOs and potential NCOs was firmly in place for the Army.

"FOLLOW ME"

NCOC 18-70, 70 COMPANY
7th Student Battalion
Graduation 12 February 1970

A graduation booklet for Ft Benning NCOC Course 18-70.
Author's collection

Project Proficiency

A soldier's attitude towards the Army and his motivation to do his best require the best possible management of these programs to secure decisions which make sense from the individual's viewpoint. These decisions must add up to sound career management development for him.
-Enlisted Grade Structure Study, July 1967

The call was out in the Army to educate noncommissioned officers. In 1963 a council of senior NCOs at Fort Dix called for a senior NCO college, and one of the main topics was NCO education in November 1966 during Sgt. Maj. of the Army Wooldridge's first Sergeants Major Conference. The Army began to look at educating noncoms in earnest. In August 1965, the Chief of Staff of the Army directed a comprehensive Enlisted Grade Structure Study.

This study, which was completed in July 1967, focused on how to establish and manage a quality-based enlisted force, and dedicated a portion for "improving the vital area of training."

The first Sergeants Major Conference hosted .Sgt. Major of the Army William O. Wooldridge, November 1966.
Photo Credit *US Army*

In response, the Deputy Chief of Staff for Personnel developed a comprehensive 5-year plan to manage career enlisted soldiers which included many far-reaching programs, such as career management fields, MOS reclassification, the Qualitative Management Program, and Force Renewal through NCO Educational Development.

The Project recommended formal leadership training designed to prepare selected career-enlisted personnel for progressive levels of duty and noted it would enhance career attractiveness and the quality of the noncommissioned officer. This study recognized that:

"*The present haphazard system of career development, as opposed to skill development, had two bad results. First, the image of the NCO as a professional, highly trained individual is difficult to foster; second, the Army's resource of intelligent enlisted men, anxious to develop as career soldiers, is inefficiently managed. The Army has extended great effort to ensure the selected development of its officers. Analagous [sic] effort should be spent in the development of the noncommissioned officers of the Army.*"

The report went on to recommend a three-level educational program, similar to officers, outlined in the February 1969 NCO

Educational Development Concept. The first of the three levels consisted of the Basic Course which was designated to produce the basic E-5 NCO. The Advanced Course was targeted to mid-grade NCOs, and the Senior Course was envisioned as a management course directed to qualifying men for senior enlisted staff positions.

The Skill Development Base program, specifically NCOCC, was selected to serve as the model for the Basic Course. Project Proficiency, to be known as the NCO Education system, was to become a reality. On the 23rd of April 1970, President Richard Nixon announced to Congress that a new national objective would be set to establish an all-volunteer force and from that the Modern Volunteer Army was born. But by mid-1971 Army Chief of Staff General William Westmoreland was unhappy with the progress of the MVA and asked then retired General Bruce C. Clarke to travel the Army and find out what could be changed to make it more attractive.

First graduating class of the Fifth Army and III Corps NCO
Academy, Mar 10, 1972.
Photo Credit *US Army*

On a visit to Fort Hood, TX, Clarke arrived in time for its NCO Academy to close its doors, a repeat of the same story at other installations.

Clarke conducted a survey and discovered that there were only four NCO Academies remaining in which to train 100,000 noncommissioned officers. In his report back to Westmoreland, Clarke lamented that "we are running an army with 95% of the NCO's untrained!" NCO academies across the nation were reopened, and Westmoreland approved the Basic and Advanced noncommissioned officer courses, and by July the first Basic course pilot began.

Some of the difficulties facing the Army of 1971 included Westmoreland's concern for leadership inadequacies. He directed the CONARC Commander to conduct a study on leadership and noted "the evident need for immediate attention by the chain of command to improving our leadership techniques to meet the Army's current challenges." He also directed the War College in Carlisle,

Pennsylvania to determine the type of leadership that would be appropriate as the Army approached the end of the draft.

In April 1970 and on the heels of a final report of the My Lai massacre and several unfavorable events led to Army Chief of Staff to desire to assess the moral and professional climate of the Army during a most turbulent period and commission a number of studies. While these studies

[H.A.S.C. No. 92-51]

HEARINGS

BEFORE THE

SPECIAL SUBCOMMITTEE ON THE UTILIZATION
OF MANPOWER IN THE MILITARY

OF THE

COMMITTEE ON ARMED SERVICES
HOUSE OF REPRESENTATIVES
NINETY-SECOND CONGRESS
FIRST AND SECOND SESSIONS

OCTOBER 13, 26, NOVEMBER 4, 19, 1971; MARCH 6, 1972

[Pages of all documents printed in behalf of the activities of the House
Committee on Armed Services are numbered cumulatively to
permit a comprehensive index at the end of the Congress. Page numbers lower than those in
this document refer to other
subjects.]

U.S. GOVERNMENT PRINTING OFFICE
WASHINGTON : 1972

Subcommittee Report of the
Utilization of Manpower in
the Military, 1972.

were going on, the Army was continually under fire. The May 1971 release of the Comptroller General's Report to Congress on the Improper

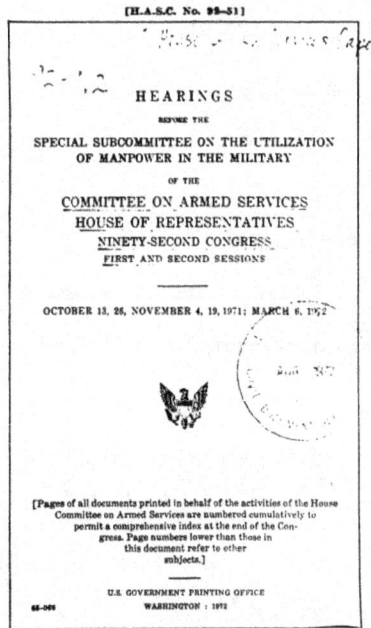

Use of Enlisted Personnel noted that the Secretary of the Army should strengthen existing policies rather than introduce latest programs or changes. That same month Westmoreland urged all the commanders of the major commands to grant their noncommissioned officers' broader authority. In his list of 14 points he asked them to "expand NCOs education through wise counseling and by affording them [NCOs] the opportunity to attend NCO Academies, NCO refresher courses, and off-duty educational programs."

Creating a Noncommissioned Officers
Education System

*The purpose of the Noncommissioned Officer Education System is to
build NCO trust and confidence, to raise tactical and technical
competence and to inculcate the essential values of the
professional Army ethic through the corps.*
Col. Kenneth Simpson and CSM Oren Bevins,
Commandant and CSM, Sgt. Maj. Academy, Oct 1989

Planning for the development of an
education system began in early 1969.
Obviously, if the NCO could be school-trained
for the jungle, then they ought to be school-
trained for the garrison, too. Westmoreland
had intended to establish a senior NCO school
in 1968, but CONARC commander General
James K. Woolnough was not enthusiastic
about the plan. Woolnough believed that
senior NCOs, like generals, needed no further
military schooling.

That was the same problem General Johnson
was earlier faced with while trying to establish

the NCO Candidate Course when CONARC commander General Paul A. Freeman and his headquarters would not accept the idea. Johnson opted to wait until Woolnough assumed command of CONARC to begin the NCO candidate program. Westmoreland would also wait until General Ralph E. Haines Jr. succeeded Woolnough at CONARC.

CONARC NCO Education and Development Study, 1971.

In July of 1970, during a lull in the NCO Candidate classes at Fort Sill, they conducted the first pilot of the Basic Course. The NCO Education Program could only begin when NCO Candidate Courses were completed because of scarce resources and the first of the Army-wide courses began in May 1971. In January of 1972 the first two Advance Courses began and that same year Chief of Staff General Creighton W. Abrams approved the establishment of the Senior NCO Course, to be located at the newly established Sergeants Major Academy at an unused airfield in El Paso, Texas. The draft ended on December 31, 1972 and the Army entered 1973 prepared to fully rely on volunteers to fill their ranks.

The three-tiered (later four, now five) noncommissioned officer education system was initially developed as an in-service program for career soldiers, specifically for those who had re-enlisted at least once. Students would attend the courses in a temporary duty status, with the sergeants major course being a permanent change of station. NCOES was established in late 1971 and phased in across the Army.

Funding was a problem, particularly with overseas soldiers and by December 1971 CONARC had to cancel nine of twelve Basic Course classes because of poor attendance. CONARC convened a NCOES conference in October and implemented incentives including promoting the top graduates, offering promotion points to graduates and mandatory quotas by CONARC.

Reserve soldiers were authorized to attend active courses, and different branches developed correspondence courses. By mid-1973 forty-one basic courses were in operation. In January 1972 the first two advance courses started, consisting only of E-7s because the Department of the Army did not maintain the files of E-6s to screen. By 1974, forty-two courses had been established through

CONARC, and in August U.S. Army Europe personnel were allowed to attend advance courses in the United States.

The Sergeants Major Course

"I feel the program of instruction is very demanding, particularly in the areas of human relations and military organization and operations."
-Msgt. Henry Caro, Excellence in Leadership
awardee, SMC Class No. 2, 1974

Department of the Army General Order 98 on July 15, 1972 authorized the U.S. Army Sergeants Major Academy. This capstone senior level course was designed to prepare selected E-8s for duty as sergeants major and command sergeants major throughout the Army. Unlike other NCOES courses it was branch immaterial and similar to courses provided to commissioned officers at senior service schools. To develop the initial curriculum a committee was formed consisting of ten command sergeants major from major commands and thirteen educational personnel from throughout the Army's schools' systems.

Some of the subjects selected included the usual courses on leadership, military organization and military management. But the sergeants major course also included topics on world affairs and human relations.

Some within the military were against NCOs studying world affairs. The subject was supported and encouraged by the first Commandant, Col. Karl Morton, and the Command Sergeant Major, then Cmd. Sgt. Maj. William G. Bainbridge (who would later go on to become Sergeant Major of the Army). It took a decision by the Army Chief of Staff, Gen. Abrams, to keep this topic in the course.

Sgt. Maj. Of the Army William G. Bainbridge.
Photo Credit *US Army*

In December of 1972 students began to arrive and on January 15, 1973, the initial Sergeants Major Course of 105 students was convened.

This first class was organized with student leaders who attended to administrative details and organizing committees. This first class also established a new tradition by conducting a dining-in and dining-out for students. These formal functions were a tradition with the officer corps but not for noncommissioned officers, and the first was held March 22, 1973.

Original Sergeants Major Academy building.
Photo Credit *US Army*

The senior course was designated the "capstone" of the noncommissioned officer education system. It consisted of over 600 hours of instruction, mostly classroom centered, using a "small group" process. This method centered on a participatory method of training in which 10 or 15 students were organized into groups and accomplished a majority of their learning by doing. This different approach let the students participate in the learning.

The Small Group Instruction (SGI) process shifted the teaching methodology from "what to think" to "how to think," and placed the learning responsibility on the student through group participation and assignments as discussion leaders. Typically, the first students were first or master sergeants with between fifteen and twenty-three years of service. As long as a soldier was not a serving sergeant major, he or she could attend.

A modern view of the NCO Leader Center of Excellence.
Photo Credit *US Army*

The Academy offered a comprehensive, professional educational environment in which each individual was offered an opportunity to broaden his knowledge and discover new fields outside his MOS. Besides the academic portion, students were offered a college electives program and received an opportunity to participate in a college degree program.

The Enlisted Personnel Management System

The NCOES complements the Enlisted Evaluation System, except that each course goes deeply into the hands-on skills required in the core *duty positions of the MOS while the corresponding test assesses knowledge across the breadth of the MOS.*
-Brig. Gen. William Patch, former Director of Enlisted Personnel, Nov. 1974

Educating NCOs would forever be different after the implementation of NCOES, but in the meantime the NCO Academy structure continued to operate in parallel as the NCOES began operations. The declared purpose of noncommissioned officer academies remained constant, an in-service leadership training opportunity to develop noncommissioned officers and specialists in fundamentals and techniques of leadership, and to offer increased career educational opportunities.

Considered a steppingstone to promotion,

NCO Academies were also used to prepare NCOs for leadership duty in all environments and to instill in them self-confidence and a sense of responsibility. At Fort Bragg, NC the academy delivered a four-week curriculum to one hundred students at a time to not only instruct, but to "raise the prestige of the of the Army's corps" of NCOs.

Academies were still established by divisions or at installations and the CONARC replacement, Training and Doctrine Command (TRADOC), approved the programs of instruction. Graduates of the basic and advance courses (NCOES) were not allowed to attend NCO academies, and academies were discouraged from being used as "pre-NCOES preparation" courses. In Europe, the Seventh Army NCO Academy officially designated that the Commandant would be an enlisted soldier, and Cmd. Sgt. Maj. Lawrence Hickey became the first in January 1972.

Sgt. Maj. Lawrence Hickey with Bob Hope in Vietnam.
Photo Credit *US Army*

As an outgrowth of Project Proficiency, the Army Chief of Staff directed that an Enlisted Personnel Management System Task Force be formed to conduct a sweeping review of enlisted personnel management. This task force was organized in January 1973 to design a career system that would challenge, develop, reward, and satisfy soldiers so well that more would want to stay for a career.

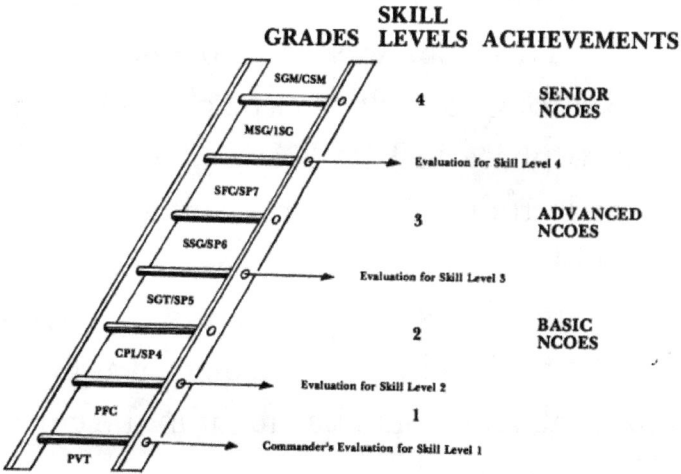

THE CAREER
ACHIEVEMENT
LADDER

GRADES	SKILL LEVELS	ACHIEVEMENTS
SGM/CSM	4	SENIOR NCOES
MSG/1SG		
	Evaluation for Skill Level 4	
SFC/SP7	3	ADVANCED NCOES
SSG/SP6		
	Evaluation for Skill Level 3	
SGT/SP5	2	BASIC NCOES
CPL/SP4		
	Evaluation for Skill Level 2	
PFC	1	
PVT	Commander's Evaluation for Skill Level 1	

EPMS Career Ladder.
Commanders Call, DA Pam 360-817, Spring 1974

It also would provide the right number of soldiers in the right grades and skills to carry out the Army's mission. It would serve to eliminate the dead-end military occupational specialties, those in which a soldier could only advance as high as sergeant.

NCOs were now allowed to merge to specialties at a higher grade in a similar career field without changing having to change jobs entirely. This plan was to implement a new EPMS through a multi-year plan. As the Army began to phase in EPMS one of the changes was the introduction of a primary level course to be added to NCOES. A 3-4-week Primary Noncommissioned Officer Course (PNCOC) would be for combat arms soldiers, was branch immaterial, and would be taught in the current NCO academies.

At the same time, the basic course was to be shortened and by 1976 to be redesigned as the Basic Noncommissioned Officer Course (BNCOC). Also, in 1976, TRADOC directed that a Primary Leadership Course (PLC) be developed to train the first line leaders in

Combat Support and Combat Service Support fields and to also be taught at the NCO academies. Soon, the Advance courses were also redesigned as the Advance Noncommissioned Officer Course (ANCOC) to support EPMS.

EPMS was implemented on October 1, 1975 and was designed to provide clear patterns of career development and promotion potential. A goal of eliminating bottlenecks for promotion was established by grouping MOSs into career management fields. EPMS quickly took over, expanded and integrated NCOES, and took the basic combat arms courses out of the service schools and placed PNCOC/PLC and BNCOC with the NCO academies. EPMS was extended to the Army Reserve the beginning in the following year.

Evolution

In some respects, training in today's Total Army is similar to training in years past. General concepts remain the same. Officers set standards, and NCOs train soldiers and small units up to those standards.

-PLDC Handbook, Army Chief of Staff
and SMA, 1989

Among other things, the EPMS plan was to tie NCOES to pay grades and promotions. Its impact on NCOES was long range and far reaching, phased to be accomplished by 1977. Pilot courses of the PNCOC were held in the summer of 1975 at Forts Carson and Campbell. The transition from NCO academy leadership courses to PNCOC, and later PLC, took well into 1978.

A separate group of courses was developed in 1976 for the combat support and combat service support NCOs that were technical in nature. The PLC for CS/CSS were mainly leadership oriented and for the most part MOS immaterial, but there were no BNCOC leadership equivalent courses. But as the combat soldiers learned job-related skills in PNCOC, the Primary Technical (PTC) was introduced to complement PLC, and the Basic Technical Courses (BTC) became the CS/CSS soldiers basic level course. These were typically conducted at the service school responsible for the management of the particular career field.

Sgt, 1st Class Goff leading a class in 1979 at the Seventh Army NCO Academy.
Photo Credit *US Army*

Over the five years that followed, NCOES would continue to undergo implementation and changes. Overseas soldiers had difficulty in attending courses in the United States and TRADOC did not want to establish PTC/BTC overseas for CS/CSS soldiers.

Basic Technical Course 76Y, 1986, Fort Lee, VA.
Photo Credit *US Army*

Attendance regularly fluctuated throughout the period due to travel fund shortages and lack of interest. By 1979 and 1980, the TRADOC commander called for a survey of BNCOC and ANCOC in conjunction with a revision of the governing regulation, AR 351-1, *Individual Military Education and Training*. A result was a Common Leader Training portion added to both courses in the early 1980s, and TRADOC announced that a new course would replace the current primary level courses.

In order to promote retention and encourage soldiers to want to stay in the Army a Cohesion and Stability Team (ARCOST) was formed and made a number of recommendations, including adding new peacetime awards. The subject of military awards first surfaced during a September 1980 news conference when Army Chief of Staff General Edward C. Meyer stated the Army was studying the creation of several military awards to recognize Soldiers' contributions to the Army during peacetime.

Shown with the numeral 4, the NCOPDR is displayed ahead of the Army Service Ribbon.
Photo Credit *US Army*

Established by the Secretary of the Army John O. Marsh Jr. on April 10, 1981 and effective August 1, 1981 was the creation of the NCO Professional Development Ribbon, to be awarded to members of the US Army, Army National Guard, and Army Reserve for successful completion of designated NCO professional development courses. In reports of the time "the NCO Academy ribbon will be awarded to enlisted soldiers upon completion of each level of the NCO Education System." It was initially reported that subsequent awards were to be designated "by an oak leaf cluster," which changed to numerals upon implementation.

On July 23, 1982 TRADOC directed that PNCOC and PLC be combined to form a Primary Leadership Development Course (PLDC). This new course was to be implemented in January 1984 and the Sergeants Major Academy would become the proponent for its development. The initial courses were conducted at Forts Leonard Wood and Polk in 1983 with much success.

When released in 1984, the new AR 351-1 mandated the establishment of Order of Merit (OML) lists at the battalion level, which would cause better attendance and decentralize control over which students were selected to attend the initial leadership training program. The Sergeants Major Academy had become the proponent for ANCOC common leader training in June 1981, and with the release of the new regulation in 1984, also assumed the "common core" for BNCOC.

PLDC graduation photo, Class 7-87.
Photo Credit *US Army*

Implementation of PLDC was nearly complete by the end of 1985, so when PTCs were abolished in that December, PLDC became the sole MOS nonspecific basic course for inservice NCO leadership training. PLDC was the NCOs first step to education, and the leadership and tactical training was aimed at the junior noncom.

One of the recommendations of a December 1985 NCO Professional Development Study group, aptly labeled the "Soldier's Study," repeated earlier recommendations to tie NCOES to promotion. The Army Chief of Staff approved the concept that NCOES be made mandatory, sequential, and progressive, and with NCOES-promotion linkage. In 1985, the small group instruction method became standard for all NCOES courses.

January 1986 began with the redesignation of BTCs as BNCOC-CA/CSS and the establishment of an Operations and Intelligence Course at the Sergeants Major Academy. This functional course would become the predecessor of the branch immaterial Battle Staff NCO Course.

A grader scoring the push-up event at the Air Defense Artillery NCO Academy.
Photo Credit *US Army*

To become effective by July 1986 and to be considered for promotion to staff sergeant a soldier had to complete PLDC, and then a prerequisite Army-wide also required that PLDC attendance was mandatory for BNCOC attendance and would go into effect that October.

Standardizing throughout the 1990s

A noncommissioned officer corps, ground in heritage, values and tradition, that embodies the warrior ethos; values perpetual learning; and is capable of leading, training, and motivating soldiers.
-The Noncommissioned Officer Corps, *Future Leader Development of Army NCOs Workshop*, 1998

The final decade of the 20th century opened with a swift victory in the Gulf, Operation Desert Storm was declared a major victory for democracy and the US led coalition in rolling back aggression in the middle east. One of the successes touted in the limited war was the demonstrated professionalism of the all-volunteer noncommissioned officer force, and how the investment s made in unit training and individual leader development were key to the successes.

The Noncommissioned Officer Education System underwent many studies, improvements, and revisions from the mid-1980s to the end of the century. A 1989 NCO

Leader Development study noted that NCOES was not completely aligned with unit levels of leadership and went on to recommend

PLDC Student leader being graded in 1992 at Camp Jackson, Korea NCO Academy.
Photo Credit *US Army*

requiring attendance to promotion; PLDC for sergeant, BNCOC for staff sergeant, ANCOC for sergeant first class, and the sergeants major course for sergeants major. At its peak in 1992, about 90,000 students had graduated from noncommissioned officer education system courses.

Some of the recommend improvements included adding rifle qualification requirements, a train-the-trainer course, and "shared" field-training exercises between other training programs such as basic combat training. Automated systems were being used to track order of merit lists and in scheduling students for training. special NCOES courses were designed for Reserve Component schools and a special task force was established to evaluate training for the Guard and Reserve. By the end of the Cold War in 1991, NCOES was an integral part of the Enlisted Personnel Management System and was widely credited for contributing to the US Army's success in the Gulf War.

The soldier of the twenty-first century required

appropriate training before promotion to the next grade level under the Select, Train, Promote (STP) system. The Sergeants Major Academy was a major force in the history of training noncommissioned officers. Not only did the Academy serve as proponent for PLDC and the common leader training of the other NCOES course, it was also responsible for the functional courses Battle Staff NCO, the First Sergeants Course, and the Command Sergeants Major Course. The stated goal of the NCOES and noncommissioned officer training was to prepare noncommissioned officers to lead and train soldiers who work and fight under their supervision and assist their assigned leaders to execute unit missions.

In utilizing modern technologies which brought the classroom to the student, desk-top computer technology and the Internet, new methods were trialed and tested to offer "distance learning" to the remotely located student. In the spring of 1995, a pilot course was conducted to teach PLDC using a video tele-training (VTT) techniques across the airwaves in an interactive session to soldiers on duty in the Sinai. And in 1997 the common leader portion of ANCOC was taught to reserve component NCOs at Fort Hood, Texas.

Delivering ANCOC via distance learning for the first time with the Texas Army National Guard.
Photo Credit *US Army*

Todd A. Weiler, the Deputy Assistant Secretary of the Army for Reserve Affairs, Mobilization, Readiness and Training, who had been instrumental in the development of the Army Distance Learning Program strongly emphasized that distance learning is a "must." and that "the future of the Army must involve distance learning." Computer based CD-ROM courses tied to a workstation assisted students in

The NCO Vision

An NCO Corps, grounded in heritage, values and tradition, that embodies the warrior ethos; values perpetual learning; and is capable of leading, training and motivating soldiers.

We must always be an NCO Corps that
- Leads by Example
- Trains from Experience
- Maintains and Enforces Standards
- Takes care of Soldiers
- Adapts to a Changing World

Effectively Counsels and Mentors Subordinates
Maintains an Outstanding Personal Appearance
Disciplined Leaders Produce Disciplined Soldiers

SMA Jack L. Tilley
12th Sergeant Major of the Army

NCO Vision.
FM 7-22.7, *The Army NCO Guide*,
December 2008

learning as modern technologies were introduced to digital classrooms. Methods for institutional training became more complicated but the need for trained NCOs remained the same.

A "Vision" for the future NCO was developed through a series of workshops at the end of the decade to establish a perpetual learning pathway for the noncommissioned officer corps to lead the NCO Corps into the twenty-first century. Other than minor changes going into the 2000s, NCO leadership training and education merely underwent a series of minor changes and improvements. But that was soon to change.

Embedding Technology

*[Classroom XXI] focuses on the leveraging of technology to use information in a variety
of ways so as to increase the Army's warfighting capability.*
-How the Army Runs: A Senior Leader Reference
Handbook, 2001- 2002

The terror attacks on America on September
11, 2001 would be the cause of a number of
significant changes to Army training and
education that continue today. Since the 1980s
the US Army had found itself involved in
short-term global operations from dealing with
despot leaders in Panama, Somalia and Iraq,
intervening between warring factions in eastern
European countries like Bosnia Herzegovina,
and Kosovo, to peacekeeping duties in the
Sinai and Macedonia, none of those would
compare to their level of involvement in a
Global War on Terrorism.

The US response was swift, first in Afghanistan
and then shifting to Iraq. Involvement in the
middle east had become a long-protracted

series of deployments and redeployments, the Army through its collections of observations of the insurgent warfare retooled its training across the force and was reflected throughout NCOES.

There was a move by Army leaders to harness technology and shift some learning remotely and virtually using digital technologies. Some of the combined results included a reduction in the amount of time a soldier spent in school, less travel, less time away from their units (and home) and resulted in a significant cost savings. Classrooms were modernized and video cameras and remote platforms were integrated for digital training facilities in what would be called *Classroom XXI.*

The Army directed the transformation of the Primary Leadership Development, Basic NCO, Advanced NCO and the Sergeants

Major courses in 2004, to what would evolve into the Warrior Leader, Advanced Leader, and Senior Leader courses and a completely redesigned Sergeants Major Course. And, NCOs were required to complete both a common core and an MOS specific phase to successfully complete a new Advanced Leader Course.

The Army announced that its Primary Leadership Development Course was to be renamed the *Warrior Leader Course,* (WLC) in October 2005. Officials noted that the new name

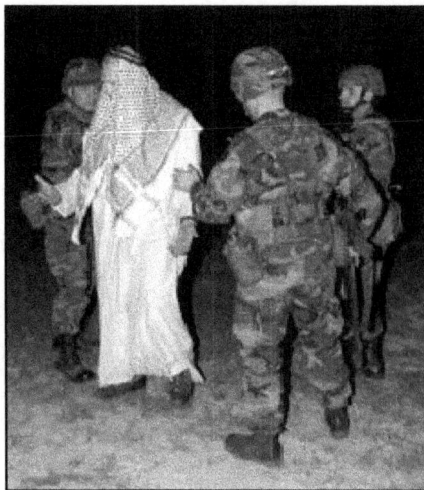

WLC students at Fort Dix NCOA.
Photo Credit *US Army*

would reflect changes made to the PLDC

curriculum and the course was redesigned to better prepare soldiers for asymmetrical warfare and would include lessons from the wars in Iraq and Afghanistan.

In reviewing change for future requirements, the *Army Leaders for the 21st Century: Final Report* was published in November 2006, which described the Army's *Review of Education, Training and Assignments for Leaders* (RETAL) report. This study looked to history to chart a path to the future and how to change the Army's leader development model for the future. The objective was to produce "pentathlete leaders" and created an

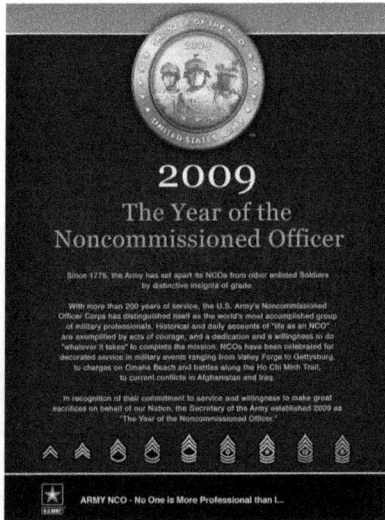

Year of the NCO decree.
2009 Army Posture Statement

Army Pentathlete Leader Model, eventually replaced by a Leadership Requirements Model and an Army Leader Development Strategy.

In 2009 as part of the second Year of the NCO the Army revamped NCOES by renaming the Basic and Advanced Noncommissioned Officer Courses to the Advanced Leader Course and the Senior Leader Course, respectively. To be completed in late 2010, the new ALC and SLC were to focus beyond what was required for an NCO at their current rank and position, but to prepare them for future requirements.

A primary purpose was to provide NCOs with training early enough in their career to affect the more complex leadership challenges they were expected to be faced with in a changing environment. The BNCOC Phase I distance learning program was transformed to

Advanced Leader Course-Common Core in October 2009.

An Institute for NCO Professional Development (INCOPD) was created by TRADOC in August 2009 to bring the various and fragmented NCO professional development programs under a single entity. The concept for the INCOPD was to integrate

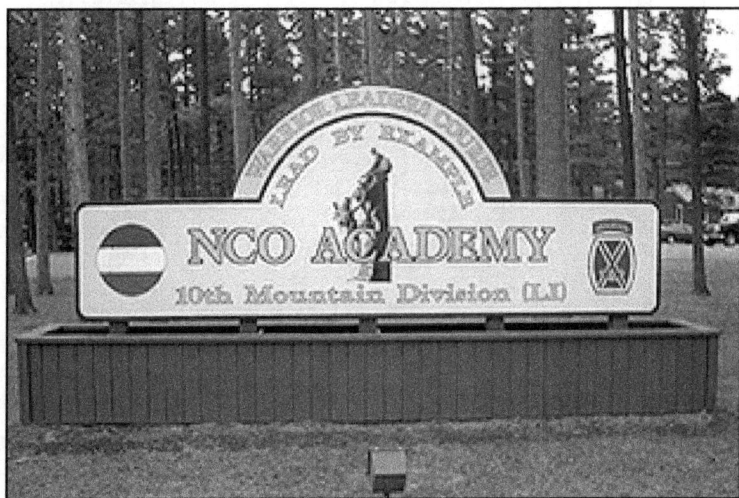

10th Mountain Division NCO Academy.
Photo Credit *US Army*

all NCO development activities and the staff reported directly to the Training and Doctrine Command Commander. TRADOC had been looking at providing more structure to enlisted leaders for self-development since the 1997 *Future Leader Development of Army NCOs* workshops and so the Sergeants Major Academy was handed the mission and created a program for multi-level self-development.

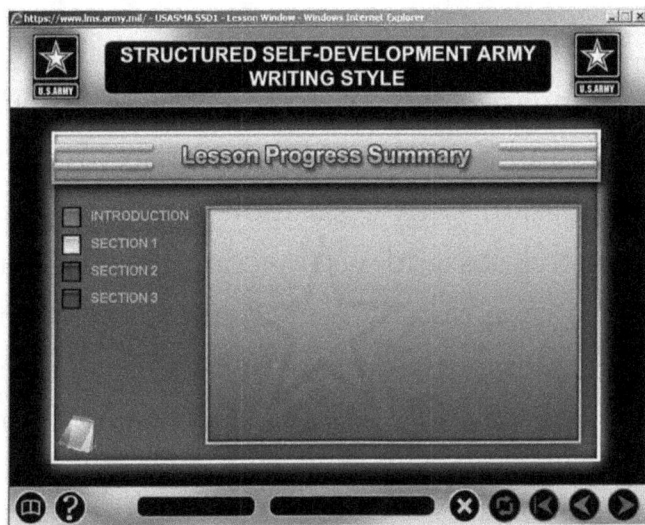

SSD I, Module 1, Army Writing Style.
SSDI screen

In October 2010 USAMA completed the development of the first level of Structured Self-Development (SSD) with the concept to provide learning opportunities between NCOES courses across a soldier's career. Beginning with SSD I as a prerequisite for the new WLC, SSD's were mandatory web-based training that built upon and added to lessons delivered in NCOES courses.

Each SSD was required to be completed prior to attendance to Warrior Leader, Senior Leader, and Sergeants Major courses. It began as a program with both individual and leader responsibilities and was accomplished on the soldier's own time while serving at their home unit. Implementation was to be phased over a three-year period as SSD's I through V, and implementation was set to be completed by mid-2013.

NCO Professional Development System

To best prepare our NCO Corps for the challenges of an uncertain future, we must fundamentally change and evolve the Noncommissioned Officer Education System into a comprehensive leader development system that links training, education, and experiences spanning the operational, institutional, and self-development learning domains.
-NCO 2020 Strategy, TRADOC, Dec. 4, 2015

Though the content and course material have changed considerably over the 45 years since NCOES was conceived, its framework had stayed constant. As a result, a roadmap for future development was created as the *NCO 2020 Strategy* established the NCO Professional Development System (NCOPDS) to absorb and expand NCOES across three focus areas: *development*, *talent management* and *stewardship*.

The belief was that education was no longer just a part of the EPMS, but an independent culmination of both the Career Map, formally known as the *Army Career Tracker*, and a Select-Train-Educate-Promote (STEP) approach.

Army Career Tracker dashboard.

Adding to the previous Select-Train-Promote model a change toward STEP was the new requirement for NCOs to attend formal education to become certified at their grade level before being qualified to advance to the next grade.

In creating the NCO 2020 Strategy, data was used from the 2011 Center for Army

Leadership Annual Survey of Army Leadership, which explored Army Professional Military Education (APME) attendance, quality, effectiveness and relevance along with the ability to transfer what was learned in the classroom to the field.

The study revealed significant shortcomings in leader development; especially in the areas of critical thinking, the ability to apply skills in the operational setting, and in problem-solving, leadership, communication, management, interpersonal, technical, and tactical skills. NCOs were losing competency in those areas, which were compounded and reduced their ability to exercise sound leadership principles.

A working group from the INCOPD and the Sergeants Major Academy came together in the spring of 2015 to examine the survey data and determine what could be done to improve the

NCO corps. The group discovered that the method of preparing NCOs through education was not aligned with contemporary doctrine and concepts because NCOES failed to provide progressive and sequential leader development, with gaps existing between the Warrior Leader Course to the Sergeants Major Course.

The group believed that existing NCOES courses focused on technical and tactical training causing gaps in developing NCO leader core competencies. This was compounded by having too many hours of directed and mandatory training, and it was discovered that the time between the Senior Leader Course and the Sergeants Major Course was upwards of seven years without any formal education. All this failed to achieve changes in behavior or affect long-term knowledge retention.

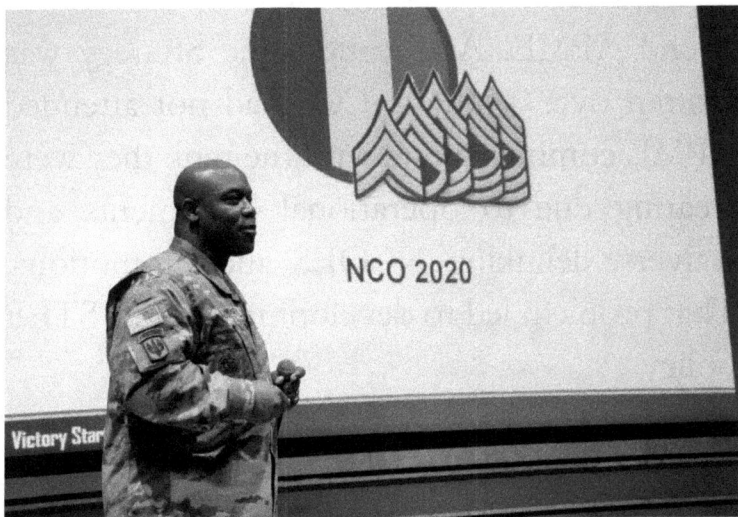

Sgt. Maj. Brian Lindsey of the INCOPD briefs USASMA Class 68 on the NCO 2020 Strategy, Dec. 2017.
Photo Credit *US Army*

This was the cornerstone in the development of the NCO 2020 Strategy, which included 45 initiatives along its three lines of effort. It described how the Army was going to develop NCOs differently, how to manage their development and how to reinforce NCO requirements. Before NCOs could begin to be developed through education, they had to attend APME. At the time the Strategy was written over 14,000 NCOs had not attended APME commensurate with the rank they were wearing due to operational deferments and waivers, delinking NCOES and promotions. That problem led to development of the STEP policy.

STEP recognized that only the best soldiers should be selected to become NCOs based on their performance, potential to lead and character. The unit had the responsibility to take that soldier and invest time and energy in further developing them to lead in organizations throughout the Army. Potential NCOs were validated by appearing in front of a board of senior NCOs, and then it became the responsibility of the proponent schools to educate them in the duties they were about to assume before promotion to the next rank.

With the July 2015 publishing of EXORD 236-15: *Army-Wide Implementation of Noncommissioned Officer Professional Development System*, the Training and Doctrine Command directed redesign of the Basic Leader Course, created Leader Core Competencies for proponent

Graduates of pilot #3 of the Master Leaders Course, Mar. 29, 2016.
Photo Credit *US Army*

Advanced Leader and Senior Leader courses, created the Master Leader Course, accredited the Sergeants Major Course as a degree-granting program, and revised all levels of structured self-development that created a revolutionary change to all six levels of NCO APME.

The Master Leader Course (MLC) was a way to "train ahead" by closing the upwards of seven-year training gap between a senior NCO's attendance at the Senior Leader Course and the Sergeant Major Academy. The two-week course was designed to educate Sgts. 1st Class in the areas of professional writing, communication skills, critical thinking, organizational and command leadership, management skills, and joint and operational level warfighting, and became mandatory in fiscal year 2019.

Highlighting the varied duties, the original Sergeants Major Academy had become it was redesignated as the U.S. Army Sergeants Major Academy at the NCO Leadership Center of Excellence. On March 21, 2018, Combined Arms Center officially designated the Sergeants Major Academy a branch campus of the Command and General Staff College allowing qualified graduates of the Sergeants Major Course to earn a Bachelor of Arts in Leadership and Workforce Development from USASMA.

Epilogue

The sergeants and corporals will be, not only how to execute with precision the manual of arms as sergeants, but, also, every thing relating to the manual of arms, as rank and file, the firings and marching.
-Winfield Scott, *Infantry tactics: or, Rules for the exercise and Manoeuvres of the United States' Infantry*, 1857. p. 19

The goal for the NCO Professional Development System was to prepare the next generation of NCOs to develop as leaders over time. It was to be done through a progressive and sequential process to incorporate training, education, and experience across all three learning domains of self, operational and institutional. The STEP framework assigned training and education responsibilities across those three learning domains for development and advancement of the professional NCO.

As the "first rung" of in-service enlisted leader training a redesigned BLC went into effect in February 2019. Created as a 22-day course and delivered at 34 active duty NCO Academies, with each starting and stopping on the same day. BLC curriculum consisted of 22 lessons divided into four phases; Foundations, Leadership, Readiness and Assessment.

Staff Sgt. Kevin Hopper, Basic Leader Course Instructor.
Photo Credit *US Army*

Designed for "getting back to basics" in order to better develop NCOs with a strong intellect, physical presence, professional competence, and moral character to serve as a role models.

The course consists of five assessments and no multiple-choice examinations. Assessments were to be done through the observation of their communication skills and written essays.

A screenshot of a Distributed Learning Course.

In 2020 the Army discontinued Structured Self Development and replaced them with the Distributed Leaders Courses. The courses consisted of six-levels of self-study and training that is linked to progressive classroom NCO career leadership courses. It was envisioned to build knowledge and skills through a sequence learning combining formal education and experiential learning.

Training had been a part of the United States Army since the days of von Steuben; however, leader development and education for noncommissioned officers has not. It was not until the Army was faced with large groups of untrained NCOs and soldiers in the aftermath of World War II that the focus initially turned to training enlisted leaders as occupation forces.

With the war in Vietnam winding down and an end to the draft expected, the Army was now faced with the challenge to entice men to consider a career in the Army to help maintain a larger standing Army. By building on the success of the NCO Academies and the NCO Candidate Course the concepts of career paths for enlisted leaders took seed.

Recognizing that military education would increase the professionalism of the noncom and improve job satisfaction it was believed a comprehensive program would sustain the Modern Volunteer Army. The investment in noncommissioned officer programs in the 1970s and 1980s were instrumental in the successes of the Army in operations Just Cause, Restore Hope, and Desert Storm, NCO successes were often attributed to the leader development programs, including a modern NCO education system.

In a review of NCO talent, an Army training command study concluded that "the development of NCOs was cumulative and sequential as a result of their military schooling, operational assignments, and self-development." With an NCO educational system beginning at the most basic leader course to a senior course known as the Sergeants Major Academy it was reported that as a system the NCOES was "producing technically competent and tactically proficient leaders" for combat.

As of this writing the US Army noncommissioned officer corps is a coveted force militarily, internationally, and across the service branches. Exuding quality and professionalism, other nations study their ways by attending the various enlisted leader training programs in their lands, or in US training centers. The Army senior course boasts an international department where students from a variety of nations attend the Sergeants Major course annually. The US Army of the 21st century requires a strong corps of enlisted leaders, and education has proven to be one of the driving factors in maintaining its edge.

Bibliography and Works Consulted

U.S. Government Documents and Publications:

Comptroller General of the United States, *Report to Congress on the Improper Use of Enlisted Personnel,* Comptroller General of the United States Washington, D.C. (May 6, 1971) pp. 1-3.

Department of the Army, Army Regulation 350-90, *Education and Training, Noncommissioned Officer Academies,* Washington, D.C. (25 Jun. 1957).

_____, Army Regulation 350-90, *Noncommissioned Officer Academies,* Washington, D.C. (19 Mar. 1968).

_____, Army Regulation 350-90, *Noncommissioned Officer Academies,* Washington, D.C. (30 Oct. 1973).

_____, Training Circular 7-22.7, *Noncommissioned Officer Guide,* Washington, DC. 1 Jan 2020, p. 1-8

_____, Enlisted Grade Structure Study, Volume 1, Part 1 *Deputy Chief of Staff for Personnel* Arlington, VA (Jul 1967) p. 1.

Friedrich Wilhelm von Steuben, Regulations for the Order and Discipline of the Troops of the United States, Part I Styner & Cist Philadelphia (1779).

Fort Benning, Georgia (53872 Sig 2), Infantry Noncommissioned Officer Candidate Course United States Infantry School (undated).

Frederick T. Abt, et. al., Leadership for the 1970's, USAWC Study of the Leadership for the Professional Soldier United States Army War College Carlisle, PA (1 Jul 1971) p. v.

Grade Structure Division, Directorate of Procurement and Distribution, Enlisted Grade Structure Management Project, Progress Report, Phase I *Office, Deputy Chief of Staff for Personnel* Washington, D.C. (25 Mar. 1969) pp. 101-103.

Harold G. Moore, Jeff M. Tuten, *Building A Volunteer Army: The Fort Ord Contribution,* Department of The Army, Washington, D.C., (1975) pp. 98, 107.

Henry E. Emerson, *Leadership for Professionals, Report of the CONARC Leadership Board,* Fort Bragg, NC (26 Jul 1971) p. 3.

Leonard Gordon, *Selection of NCO Leaders Status Report,* 30 June 1962 U.S. Army Personnel Research Office Arlington, VA (Jun. 1962) p. 1.

Noncommissioned Officer Leader Development Task Force, Action Plan United States Army Training and Doctrine Command, (Jun. 1989).

Paul D. Hood, Research on the Training of Noncommissioned Officers, Progress Report: NCO I U.S. Army Leadership Human Research Unit Monterey, CA (Jul. 1960) pp. 1-15.

_____, Implementation and Utilization of the Leader Preparation Program U.S. Army Leadership Human Research Unit Monterey, CA (Mar. 1967) p. 4.

_____, Research on the Training of Noncommissioned Officers: A Summary Report of Pilot Studies U.S. Army Leadership Human Research Unit Monterey, CA (Dec. 1965) p. 4.

_____, Morris Showel, and Edward C. Stewart, Evaluation of Three Experimental Systems for Noncommissioned Officer Training U.S. Army Leadership Human Research Unit Monterey, CA (Sep. 1967).

Richard P. Kern and Paul D. Hood, The Effect on Training and Evaluation of Review for Proficiency Testing U.S. Army Training Center Human Research Unit Presidio of Monterey, CA (Aug. 1964).

Training and Doctrine Command, NCO 2020 Strategy: NCOs Operating in a Complex World, (Dec. 4, 2015) TRADOC.

Training and Doctrine Command Regulation 351-10, Institutional Leader Education and Training Fort Monroe, VA (May 1997). Chief Signal Officer, Annual Report of the Chief Signal Officer to the Secretary of War for the Year 1872 Washington, D.C. (1873) p. 4.

U.S. Army, 1946-1951: Mobility, Vigilance, Justice *United States Constabulary,* (1951).

War Department, Annual Report 1907: Vol. I (GPO: Washington, D.C., 1907), p. 83.

Historical Reports:

Arnold G. Fisch, Jr. and Robert K. Wright, Jr., The Story of the Noncommissioned Officer Corps *Center of Military History* Washington, D.C. (June 1989).

David A. Clary and Joseph W. A. Whitehorne, The Inspector Generals of the United States Army 1777-1903 *Office of the Inspector General and Center of Military History* Washington, D.C. (1987) pp. 38-39.

James M. Snyder, The Establishment and Operations of the United States Constabulary Historical Sub-section G3, United States Constabulary Bamberg, GE (1947) p. 71.

John B. Wilson, Maneuver and Firepower: The Evolution of Divisions and separate Brigades Army Lineage Series, U.S. Army Center of Military

History, Washington, DC (1998).

Mark Gillespie, et al., The Sergeants Major of the Army *U.S. Army Center of Military History,* Washington, D.C. (1995).

Morris J. Macgregor Jr., American Military History Army Historical Series, Office of the Chief of Military Washington, D.C. (1989) p. 108.

Office of the Chief Historian, European Command, The United States Constabulary *Occupation Forces in Europe Series: 1945-1946,* Frankfurt Am-Main, GE (1947).

Oliver J. Frederiksen, The American Military Occupation of Germany, 1945-1953 Historical Division, Headquarters, U.S. Army Europe, Darmstadt, GE (1953) p. 50.

Paul T. Scheips, American Military History, Darkness and Lights. The Interwar Years 1865-1898 *Army Historical Series, Office of the Chief of Military* Washington, D.C. (1989) p. 290.

Periodicals:

A. F. Irzyk, Mobility, Vigilance, Justice - A Saga of the Constabulary (Mar. 1947) Military Review. p.19.

Alfred Hahn, Selection of Non-Commissioned Officers (Sep. 1923).

Armor, Editor, Armor NCO Candidate Course (Jan-Feb 1968) p. 53. Editor, NCO Prep Course (Mar-Apr 1969) p. 50.

Armored Cavalry Journal, SFC Philip C. Wharton and SGT Frank G. Mangin, Jr., The Armored School's Enlisted Leaders Program Richmond, VA (May-Jun. 1949) pp. 52-54.

Armored Sentinel, First NCO Academy Class Graduates (Mar. 17, 1972) Fort Hood, TX. p. 3.

Army and Navy Journal, Pershing Recommends Noncom Training (Jun. 8, 1918) p. 1567.

Army Times, Times Staff Writer, New Leadership Urged (5 May 1971) p. 3.

Bennie L. Fagan, Noncommissioned Officers Education System (Jan-Feb 1971) pp. 16-17.

Brenda Benner, Distance Learning comes to NCOES (Winter/Fall 1997) NCO Journal.

Bruce C. Clarke, U.S. Constabulary Builds an NCO Academy (May-Jun. 1950) pp. 36-38.

D. Steinmeier, The Constabulary Moves Fast (Nov 1947) Army

Information Digest. pp. 7-16.

Daniel K. Elder, Noncommissioned Officer Education System Celebrates 30 Years The US Army. Chevron.

_____, and David Davenport, Step By Step: NCO Training Has Evolved Since Army's Creation, (Mar 2018) Army Magazine.

Ernest N. Harmon U.S. Constabulary (Sep. Oct 1947) Army. p. 16.

Fred A. Darden, More Time for Sergeants (Dec 1960, Jan-Feb 1961) Infantry. p. 20.

Henry A. Finch, Increasing the Prestige of Noncommissioned Officers (Jan. 1920) p. 554.

H. P. Rand, A Progress Report on the United States Constabulary (Oct. 1949) pp. 30-38

Jimmie Bradshaw, NCO Vision (Spring 98) NCO Journal.

J. J. O'Hare, Planning the Enlisted Career Program (Nov. 1948). pp. 42-44.

John Dobbs, The CSM Academy, (Jan-Feb 1974) Infantry. pp. 37-39.

Kenneth W. Simpson and CSM Oren L. Bevins, NCOES Instills Professionalism at Every NCO Level (Oct 89) Army. p.182.

Larry H. Ingraham, Fear and Loathing in the Barracks? And the Heart of Leadership (Dec. 1988) Parameters. pp. 75-80.

Mark M. Boatner III, School for Noncoms (Aug. 1947) Infantry Journal. pp. 17-22.

Melvin Zais, The New NCO (May 1968), Army. pp. 72-76.

Robert L. Ruhl, NCOC (May-Jun 1969), pp.32-39.

R. S. Bratton, Noncommissioned Officers' Training School (Apr. 1922) pp. 429-432.

Robert Bouilly, Rise of a Professional NCO Corps (Summer 1995). pp. 36-38.

Stonnie D. Vaughan, EPMS, Career Management for Professional Soldiers (May 1974) Soldiers. pp. 15-19

Walden F. Woodward, School for Noncoms (Apr. 1947) p. 79.

William L. Hauser, EPMS (Jan-Feb 1974) pp. 26-28.

William A. Patch, Professional Development for Today's NCO (Nov 1974) pp. 15-20.

Published Works:

E. N. Harmon, *Combat Commander: Autobiography of a Soldier* Prentice-Hall Englewood Cliffs, NJ (1970).

James A. Moss, *Manual of Military Training* George Banta Publishing

Company Menasha WI (1914).

John D. Winkler, et. al., *Future Leader Development of Army Noncommissioned Officers Workshop Results* Arroyo Center, Rand (1998).

L.R. Arms, *A History of the NCO* (March 2007) U.S. Army Sergeants Major Academy, Fort Bliss, TX. (undated) p. 44.

_____, *A Short History of the NCO* U.S. Army Sergeants Major Academy Fort Bliss, TX (undated) p. 44.

Lewis Sorley, *Honorable Warrior: General Harold K. Johnson and the Ethics of Command* University Press of Kansas, Lawrence, KS (1998).

William G. Bainbridge, *Top Sergeant. The Life and Times of SMA William G. Bainbridge* Fawcett Columbine New York (Nov. 25, 1997). p.157.

William Donohue Ellis, Clarke of St. Vith, Dillon Liederbach, Inc. Cleveland, OH (1974) p. 163. Ernest F. Fisher, Guardians of the Republic Ballantine Books New York (1994) p. 292.

Oral History:

Maj. Robert L. Keeley, *Exit Interview with LTC Douglas S. Smith, Commander, 2d Bn, 47th Infantry, 9th ID*, 19th Military History Detachment Bien Phuoc, RVN (1 Jul. 1969) pp. 19-23.

Col. Francis B. Kish, Interview of Bruce C. Clarke, Gen., USA Ret. *The Bruce C. Clarke Papers, Vol. I, The U.S. Army Military History Institute* Carlisle Barracks, PA (1982) pp. 117-120.

Col. Charles S. Stoder, Interview of Isaac D. White, GEN., USA Ret. *The Isaac D. White Papers, Vol. III*, The U.S. Army Military History Institute, Carlisle Barracks, PA (1978) pp. 383-384.

Unpublished Works:

Daniel K Elder, Instant NCO, *Unpublished manuscript* Killeen, TX (2020)

John K. D'Amato, *CONT/ED Untitled Research Paper* (author's collection) (Sep 89) p. 3-3-3.

Robert Bouilly, Twenty Years of NCOES *Unpublished staff paper* (author's collection) US Army Sergeants Major Academy Fort Bliss, TX (11 May 1992) p. 5.

Robert R. McCord, History of the Seventh United States Army Noncommissioned Officer Academy *unpublished staff paper* (author's collection) Seventh Army Noncommissioned Officer Academy (1999).

Steve Ball, History of the US Army Noncommissioned Officer

Education System TRADOC Leaders Development Branch *unpublished staff paper* (author's collection) (Feb 1998).

Steven Chase, Noncommissioned Officer Education System *Unpublished research paper* (author's collection) US Army Sergeants Major Academy Fort Bliss, TX (Sep 98).

Miscellaneous:

Brian A. Libby, *The United States Constabulary in* Germany Indiana Military History Journal 13, (Oct 1988) drawn from dissertation.

Cornelius J. Shaffer, *2nd Constabulary Brigade Organization Day Mobility, Justice, Vigilance* Munchen, GE, (1951) p. 51.

Danielle O'Donnell, USASMA is now a Branch Campus under CGSC. Army.mil, (Jun. 12, 2019).

_____, Out with the old and in with the new: BLC is FOC, (Jan. 30, 2019) TRADOC.

David Crozier, Structured Self-Development and Advanced Leaders Course – Common Core, (Jul. 24, 2013) NCO Academy.

Henry C. Newton Papers (Folder) *Constabulary School History,* U.S. Army Military History Institute, Carlisle Barracks, PA

John Lossing Benson, The Pictorial Field-Book of the Revolution. 2 vols. New York: Harper, 1851–52., Public Domain, https://commons.wikimedia.org/w/index.php?curid=2262817

N.C.O.C. Locator, *Follow Me* Online document http://ncolocator.org (undated)

Noncommissioned Officer Museum Association Fort Bliss, TX (Winter 1998-1999) NCO Journal. pp. 6-7.

Peter F. Ramsberger, HumRRO, The First 50-years. HumRRO. (undated). p. 2.

Structured Self Development, TRADOC for STAND-TO! (Dec. 1, 2010)

U.S. Army Sergeants Major Academy, *History of USASMA and Fort Bliss* Online document
http://usasma.bliss.army.mil/website/a_co/r_smc/sect5.htm
(undated).

Educating Noncommissioned Officers
Daniel K. Elder

Daniel K. Elder entered the Army in Dec 1981 and served in a variety of enlisted leadership positions from squad leader to senior enlisted advisor of a 4-star Army Command. Assignments included duties as mechanic, recovery vehicle operator, squad leader, motor sergeant, senior instructor, senior drill sergeant, first sergeant, and on the staff of the Sergeants Major Academy as a lesson developer, culminating with 11 years as a Command Sergeant Major from battalion to Army major command headquarters.

He served the 2-star command of the 13th Corps Support Command at Fort Hood, TX. before his selection as the 12th command sergeant major of the U.S. Army Materiel Command, Fort Belvoir, Va. His overseas assignments include three tours to Germany and deployments to OPERATION Joint Endeavor in Bosnia-Herzegovina and Croatia and OPERATION Iraqi Freedom in Iraq.

His awards and decorations include the Distinguished Service Medal, Legion of Merit with one oak leaf cluster, the Bronze Star Medal, the Meritorious Service Medal with three oak leaf clusters; the Army Commendation medal with five oak leaf clusters; the Army Achievement Medal with six oak leaf clusters; the Armed Forces Expeditionary Medal; the Iraq Campaign Medal; the NATO Medal; the Mechanic's Badge; the

Drill Sergeant Identification Badge, and other service medals and ribbons.

Elder's military and civilian education include: Primary Leadership Development and the Primary Technical Courses; Basic and Advanced Noncommissioned Officer Courses; Instructor and Small Group Instructor Courses; Battle Staff NCO Course; Drill Sergeant School; and the Garrison Sergeants Major Course. He is a graduate of the Sergeants Major course, the Command Sergeants Major course, the Command Sergeants Major Force Management course, and the KEYSTONE Command Senior Enlisted Leader course. He was awarded a Master of Science degree in Corporate & Organizational Communication from Northeastern University, Boston, MA, and a bachelor's degree in Business Administration from Touro College, New York, NY.

Selected as the first Senior Enlisted Fellow for the Association of the United States Army office of Educations and Programs, he serves the Speakers Bureau and as a writer. A military historian, Dan has authored and edited numerous books including "Soldier for Life," the "NCO Guide (10th Ed.)" and "Sergeants Major of the Army (2nd/3rd Ed.)." A 4-time hall of fame member, he has been inducted to the U.S. Army Sergeants Major Academy "Wall of Honor" and the Halls of Fame for the Army Materiel Command, the Ordnance Regiment, and the 13th Sustainment Command. He was a 2018 General and Mrs. Matthew B. Ridgway Military History Research Grant awardee

Elder is a certified professional coach by the International Coach Federation and a volunteer with Rotary International, president of the Temple Military Affairs Committee, and a life member of the Association of the United States Army and the Veterans of Foreign Wars.

Notes:

Visit the Center for Advanced Studies of the US Army
Noncommissioned Officer at:

http://www.NCOHistory.com

Contact the author via email at:

csm@danelder.com